Guidelines for Conducting

California Standardized Regulatory Impact Assessments

DAVID METZ · BENJAMIN M. MILLER · MELISSA KAY DILIBERTI · WEILONG KONG

Sponsored by the California Department of Industrial Relations

For more information on this publication, visit **www.rand.org/t/RRA1386-1**.

About RAND

The RAND Corporation is a research organization that develops solutions to public policy challenges to help make communities throughout the world safer and more secure, healthier and more prosperous. RAND is nonprofit, nonpartisan, and committed to the public interest. To learn more about RAND, visit www.rand.org.

Research Integrity

Our mission to help improve policy and decisionmaking through research and analysis is enabled through our core values of quality and objectivity and our unwavering commitment to the highest level of integrity and ethical behavior. To help ensure our research and analysis are rigorous, objective, and nonpartisan, we subject our research publications to a robust and exacting quality-assurance process; avoid both the appearance and reality of financial and other conflicts of interest through staff training, project screening, and a policy of mandatory disclosure; and pursue transparency in our research engagements through our commitment to the open publication of our research findings and recommendations, disclosure of the source of funding of published research, and policies to ensure intellectual independence. For more information, visit www.rand.org/about/research-integrity.

RAND's publications do not necessarily reflect the opinions of its research clients and sponsors.

Published by the RAND Corporation, Santa Monica, Calif.
© 2024 RAND Corporation
RAND® is a registered trademark.

Library of Congress Cataloging-in-Publication Data is available for this publication.

ISBN: 978-1-9774-1255-3

Cover: Donald Walker/Alamy Stock; pro/Adobe Stock

Limited Print and Electronic Distribution Rights

About This Report

This report provides an overview of the California rulemaking process for executive branch agencies and guidelines for conducting standardized regulatory impact assessments (SRIAs) for major regulations. The RAND Corporation was contracted by the California Department of Industrial Relations (DIR) to review a selection of existing SRIAs to provide guidance on economic impact assessment methodologies and technical approaches for assessing the impacts of future proposed major regulations. This review is one element of RAND's work supporting DIR in the development of SRIAs and related analyses. Consistent with RAND's mission to improve policymaking and decisionmaking through research and analysis, it was determined that this review might be valuable as a public document, particularly if formatted as a broader guide for California agencies on how to conduct a SRIA.

RAND Education and Labor

This study was undertaken by RAND Education and Labor, a division of the RAND Corporation that conducts research on early childhood through postsecondary education programs, workforce development, and programs and policies affecting workers, entrepreneurship, and financial literacy and decisionmaking. This study was sponsored by the California DIR, which protects and improves the health, safety, and economic well-being of over 18 million wage earners and helps their employers comply with state labor laws.

More information about RAND can be found at www.rand.org. Questions about this report should be directed to dmetz@rand.org, and questions about RAND Education and Labor should be directed to educationandlabor@rand.org.

Justice Policy Program

RAND Social and Economic Well-Being is a division of the RAND Corporation that seeks to actively improve the health and social and economic well-being of populations and communities throughout the world. This research was conducted in the Justice Policy Program within RAND Social and Economic Well-Being. The program focuses on such topics as access to justice, policing, corrections, drug policy, and court system reform, as well as other policy concerns pertaining to public safety and criminal and civil justice. For more information, email justicepolicy@rand.org.

Acknowledgments

This report has benefited from insight, support, and feedback from several individuals and organizations. Eric Berg and Susan Eckhardt of the California Division of Occupational Safety and Health (Cal/OSHA) provided insightful state agency perspectives and helpful information on DIR's regulatory processes. The California Department of Finance provided valuable informal instruction and formal assistance (e.g., publicly available comment letters on draft SRIAs) on developing SRIAs in various communications with RAND and DIR staff. Jennifer Spore (Office of the Director, Research Unit at DIR) provided a particularly useful and detailed review of this report. We emphasize that the guidance and recommendations provided in this report are ours alone and do not necessarily reflect guidance or opinions on the part of any California agency or official. We thank our colleague Aaron Strong for input on our review of macroeconomic models for use in developing SRIAs. We also thank our two reviewers, Lynn Karoly at RAND and Lisa Robinson of the Center for Health Decision Science at the Harvard T. H. Chan School of Public Health, for their thoughtful reviews of this report.

Contents

Figures and Tables

Figures

Tables

Guidelines for Conducting California Standardized Regulatory Impact Assessments

What Is a Standardized Regulatory Impact Assessment?

The California Administrative Procedure Act (APA) established rulemaking standards and procedures for state agencies to provide the public with a meaningful opportunity to participate in the adoption of state regulations and create a rulemaking record for the California Office of Administrative Law (OAL) and judicial review (California Government Code, Chapter 3.5, Sections 11340–11365). Under California law, state agencies must conduct an analysis of the potential economic impacts of *all* regulations.[1] Specifically, "a state agency proposing to adopt, amend, or repeal any administrative regulation shall assess the potential for adverse economic impact on California business enterprises and individuals" (California Government Code, Chapter 3.5, Section 11346.3).

California Senate Bill (SB) 617 (California Senate, 2011) introduced a new requirement—as of November 1, 2013—for state agencies to conduct a standardized regulatory impact assessment (SRIA) for any regulatory action that will have an economic impact exceeding $50 million (California Government Code, Chapter 3.5, Section 11342.548). A SRIA provides a more extensive analysis of the potential benefits and costs attributed to new or changed regulatory policies, which—to the extent feasible—are expressed in

---

[1] Benefit-cost analysis is one of the primary tools used in regulatory analysis to anticipate and evaluate the likely consequences of rules (U.S. Office of Management and Budget [OMB], 2003). A SRIA also requires an assessment of macroeconomic impacts (e.g., impacts on jobs and business activity, fiscal impacts across the state). In this report, we use the term *economic analysis* to refer generally to a benefit-cost analysis and other assessments of economic impacts.

quantitative and monetary terms, in addition to considering nonmonetized impacts. The general components of a SRIA include

- a statement of the need for the proposed regulation
- a description of a baseline that reflects the anticipated behavior of individuals and businesses in the absence of the proposed regulation
- an assessment of direct benefits for and costs to businesses, individuals, and government agencies
- a detailed assessment of statewide impacts, including indirect impacts
- a comparison of proposed regulatory alternatives with an established baseline
- an assessment of distributional impacts as a result of significant differences in how the impacts of the proposed regulation accrue to different groups or over time (California Code of Regulations, Title 1, Division 3, Chapter 1, Section 2003).

SB 617 also expanded the oversight role of the California Department of Finance (DOF) (California Senate, 2011). Pursuant to California Government Code Section 11346.36, DOF adopted regulations for state agencies that conduct SRIAs. DOF's guidance is codified in the California Code of Regulations (California Code of Regulations, Title 1, Division 3, Chapter 1, Section 2003) and the State Administrative Manual (California Department of General Services, 2014).[2] For example, the SRIA must include a description of the economic impact method and approach, specific categories of businesses and individuals who will be affected by the proposed regulation, inputs into and outputs from the assessment of the economic impact, and the agency's interpretation of the results of the assessment (California Code of Regulations, Title 1, Division 3, Chapter 1, Section 2002[b][2]).[3] DOF reviews SRIAs before a major regulation is proposed and provides comments on whether the economic methodology is generally appropriate and

---

[2] DOF's instructions and code citations appear in Sections 6600–6616 of the State Administrative Manual, which pertain to the economic and fiscal impact statement (for major, nonmajor, and emergency regulations) and SRIAs (for major regulations).

[3] DOF's Standard Form 399 indicates that nonprofits should be included in the description of the types of businesses affected by the proposed regulation.

follows the relevant guidance. Agencies must include a summary of DOF's comments and agency responses when the initial rulemaking package is submitted to OAL, although agencies are not required by law to update the SRIA to incorporate these comments. DOF also generally provides technical assistance to agencies during the preparation of SRIAs.

The Purpose of This Report

The main objective of this report is to provide a concise and high-level guidebook for those individuals tasked with conducting, overseeing, or reviewing SRIAs, thereby improving the quality, usefulness, and timeliness of future SRIAs. In this report, we describe the purpose of a SRIA, when state agencies are required to conduct a SRIA, the overall rulemaking process for major regulations, the standard elements of a SRIA, high-level descriptions of the methods and tools that can be used, and commonly required revisions. We also provide a general outline for a SRIA in Appendix B. This report is not intended to provide detailed analytic guidance or training or to serve as a critical evaluation of the SRIA process or past SRIAs; where such information is included, its purpose is to provide context to those who are now facing similar decisions. We do provide broad recommendations for actions that might further promote rulemaking transparency and the credible identification of cost-effective regulatory actions.

This report is one element of the RAND Corporation's work supporting the California Department of Industrial Relations (DIR) in the development of SRIAs and related analyses. However, we have written this report anticipating that the contents might be useful to a broader audience, including California state agencies, policy analysts, various stakeholders affected by regulations, and decisionmakers who, in part, rely on economic analyses to inform regulatory policy. Agencies that have limited experience with SRIAs might find this report to be a useful first-stop guidebook. Because of general nature of its content, this report might also be of interest to regulatory economists who are interested in the discussion of methods, journalists who follow the development of major regulations in California, and state agencies in other states that follow similar administrative processes.

Research Methods

This report is based on a legal review of California state laws, regulations, and guidance documents; a high-level review of both academic and gray literature on best practices for regulatory analyses; informal discussions with DOF staff; and the learned experience of the authors in conducting SRIAs and supporting economic analyses for proposed major regulations in California. Our research was also informed by a review of completed SRIAs submitted to DOF between 2014 and 2022 to broadly characterize the variety of economic approaches employed, as well as a review of publicly available comment letters from the Chief Economist of the DOF relating to those analyses to identify common issues of concern in conducting SRIAs. Additional information related to California administrative processes was collected from the websites of DOF and OAL.

Standardized Regulatory Impact Assessment Submissions, 2014–2022

More than 200 California state agencies have been delegated authority by the California legislature to promulgate regulations implementing state law. All regulations must be approved by OAL first. Although hundreds of regulations are submitted to OAL each year—these include regular rules, emergency rules, and other minor technical adjustments that are not required to go through the full rulemaking process—generally fewer than a dozen each year are considered major regulations (Taylor, 2017). Between 2014 and 2022, state agencies proposed and conducted SRIAs for 77 major regulations, excluding withdrawn or resubmitted regulations, that covered a wide variety of topics (DOF, undated). However, there is no discernable trend in the rate at which agencies have proposed major regulations over time. Figure 1 shows the number of SRIAs submitted to OAL each year from 2014 to 2022.

Figure 2 shows the total number of SRIAs submitted to DOF and OAL by state agency. Twenty-two state agencies proposed a major regulation and submitted at least one SRIA from 2014 and 2022. Of those agencies, 18 conducted three or fewer SRIAs. Most California agencies have conducted no

FIGURE 1

Number of Major Regulations Submitted to OAL, by Year, 2014 to 2022

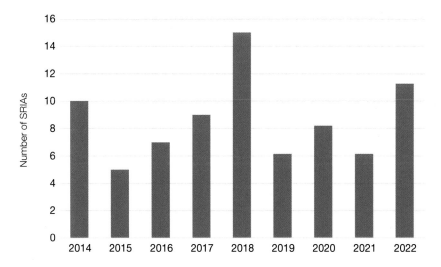

SOURCE: Features information from DOF, undated.

SRIAs as of this writing in late 2023. Some agencies have produced a relatively large number of SRIAs. The California Air Resources Board has the largest portfolio of major regulations; it conducted 28 SRIAs from 2014 to 2022, an average of more than three SRIAS per year. DIR has proposed the next most major regulations, completing seven SRIAs in that period.

Overview of the California Rulemaking Process

In general, the California legislature passes laws directing state agencies to promulgate policies that seek to achieve certain objectives but does not specify how those policies should be implemented. Through a standard rulemaking process, agencies develop more-detailed regulatory proposals in accordance with statutes to evaluate policy options, analyze the potential economic impacts of those proposals, and identify approaches that achieve the legislature's policy objectives in the most cost-effective manner. The California APA established a rulemaking process for state agencies to ensure

FIGURE 2

Number of Major Regulations Submitted to OAL, by State Agency, 2014 to 2022

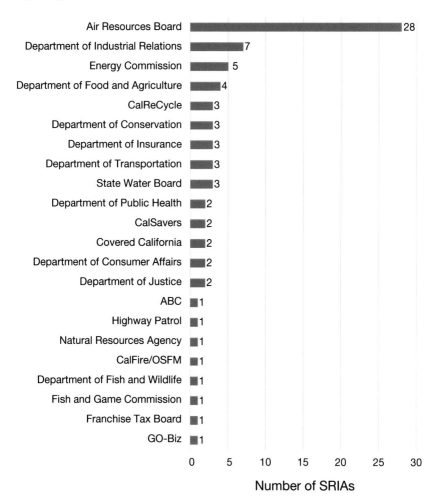

SOURCE: Features information from DOF, undated.

NOTE: ABC = Department of Alcoholic Beverage Control; CalReCycle = California Department of Resources Recycling and Recovery; CalFire = California Department of Forestry and Fire Protection; CalSavers = CalSavers Retirement Savings Board; GO-Biz = the Governor's Office of Business and Economic Development; OSFM = Office of the State Fire Marshal; State Water Board = California State Water Resources Control Board.

that proposed regulations are efficient means of implementing policy decisions in the least burdensome manner and that the regulations are "clear, necessary, and legally valid" (California Government Code, Chapter 3.5, Section 11346.3; OAL, undated-c). In accordance with the APA, an agency that plans to promulgate a new (or modified) regulation must provide justification for the proposed regulation and allow the general public meaningful opportunity to provide input before any regulation is adopted. State regulations must be adopted in compliance with regulations adopted by OAL (California Code of Regulations, Title 1, Division 1, Chapter 2, Section 280).

Determining Whether a Standardized Regulatory Impact Assessment Is Required

SB 617 (California Senate, 2011) amended the APA to establish a new process for analyzing and reviewing major regulations. Specifically, the bill requires a state agency to prepare a SRIA for any major regulation proposed on or after November 1, 2013, and submit it to DOF for review and comment. The California Code of Regulations defines a *major regulation* as

> any proposed rulemaking action adopting, amending or repealing a regulation subject to review by OAL that will have an economic impact on California business enterprises and individuals in an amount exceeding fifty million dollars ($50,000,000) in any 12-month period between the date the major regulation is estimated to be filed with the Secretary of State through 12 months after the major regulation is estimated to be fully implemented (as estimated by the agency), computed without regard to any offsetting benefits or costs that might result directly or indirectly from that adoption, amendment or repeal. (California Code of Regulations, Title 1, Division 3, Chapter 1, Section 2000)

In this context, the *economic impact* is the sum of the absolute value of all benefits and costs of the proposed regulatory action, as opposed to the net benefits (i.e., benefits minus costs). In other words, a proposed rulemaking action that is anticipated to produce $26 million in benefits and $25 million in costs would have an economic impact of $51 million and thus be considered a major regulation; the net benefits of the proposed rulemaking action would be only $1 million. This calculation includes direct impacts of the regulation, such as costs and benefits incurred by workers, employees, or customers of the regulated industry as an immediate result of changes mandated by the regulation. This calculation should also include secondary or indirect impacts of the regulation, such as changes in local tax revenue or changes in the revenues of industries that are not the immediate target of new regulatory requirements.

Although a SRIA is not required for nonmajor regulations, state agencies must still complete an economic and fiscal impact statement (Standard Form 399) that requires an analysis of the potential economic impacts of the regulation (California Government Code, Chapter 3.5, Section 11346.3). However, many of the statutory requirements for nonmajor regulations are similar to those for major regulations. For example, for all regulations, agencies are required to adopt the most cost-effective regulatory approach to address an existing problem and estimate impacts on jobs and businesses. DOF generally provides less guidance and oversight for nonmajor regulations; much of the review focuses on fiscal effects related to state and local government (Taylor, 2017). Table 1 summarizes key differences between the economic impact analysis requirements for major and nonmajor or emergency regulations.

TABLE 1

Comparison of Requirements for Major Regulations and Nonmajor or Emergency Regulations

Requirement	Nonmajor or Emergency Regulation	Major Regulation
Initial Rulemaking Package to OAL	• Notice of Proposed Action (Standard Form 400) • Express Terms • Initial Statement of Reasons/Finding of Emergency • Economic and Fiscal Impact Statement (Standard Form 399)	• Notice of Proposed Action (Standard Form 400) • Express Terms • Initial Statement of Reasons • Economic and Fiscal Impact Statement (Standard Form 399) • SRIA • DOF Comments on SRIA • Agency Response to DOF Comments

Table 1—Continued

Requirement	Nonmajor or Emergency Regulation	Major Regulation
Required Economic Impact Analysis Capabilities	"(2) The state agency, before submitting a proposal to adopt, amend, or repeal a regulation to the office, shall consider the proposal's impact on business, with consideration of industries affected including the ability of California businesses to compete with businesses in other states. (3)(b)(1) A state agency proposing to adopt, amend, or repeal a regulation that is not a major regulation or that is a major regulation proposed before November 1, 2013, shall prepare an economic impact assessment that assesses whether and to what extent it will affect the following: (A) The creation or elimination of jobs within the state. (B) The creation of new businesses or the elimination of existing businesses within the state. (C) The expansion of businesses currently doing business within the state. (D) The benefits of the regulation to the health and welfare of California residents, worker safety, and the state's environment."[a]	"(1) Can estimate the total economic effects of changes due to regulatory policies over a multi-year time period. (2) Can generate California economic variable estimates such as personal income, employment by economic sector, exports and imports, and gross state product, based on inter-industry relationships that are equivalent in structure to the Regional Industry Modeling System published by the Bureau of Economic Analysis. (3) Can produce (to the extent possible) quantitative estimates of economic variables that address or facilitate the quantitative or qualitative estimation of the following: (A) The creation or elimination of jobs within the state. (B) The creation of new businesses or the elimination of existing businesses within the state. (C) The competitive advantages or disadvantages for businesses currently doing business within the state. (D) The increase or decrease of investment in the state. (E) The incentives for innovation in products, materials, or processes. (F) The benefits of the regulations, including, but not limited to, benefits to the health, safety, and welfare of California residents, worker safety, and the state's environment and quality of life, among any other benefits identified by the agency."[b]

SOURCE: Features information from California Government Code, Chapter 3.5, Section 11346; and OAL, undated-b.

[a] California Government Code, Chapter 3.5, Section 11346.3.

[b] California Code of Regulations, Title 1, Division 3, Chapter 1, Section 2003(a).

Steps of the Rulemaking Process

The OAL outlines the regular administrative rulemaking process in California (OAL, undated-b). Proposed regulations advance through a standard process before they are finalized and approved, as shown in Figure 3. The process is initiated when a state agency identifies a problem that it intends to address through regulation. The rulemaking agency begins by gathering research and consulting with experts and stakeholders to inform the initial development of the regulation. If the agency determines that the regulatory action constitutes a major regulation, it must notify DOF of the intent to promulgate a major regulation (through a DF-130 form) and develop a SRIA to assess the potential economic impacts and compare the benefits and costs of regulatory alternatives. As an optional next step, the rulemaking agency may submit a draft SRIA to DOF, which can serve an informal advisory role to provide comment on the preliminary analysis prior to the formal submission of the rulemaking package. The agency must then formally submit the SRIA and a summary (DF-131 form) to DOF; this form provides a qualitative description of the SRIA methodology and a quantitative summary of the findings.

Within 30 days, DOF reviews the SRIA and issues a formal comment letter to the rulemaking agency. DOF might identify missing components or request additional information, such as required analyses or an explanation of the basis for key assumptions (California Government Code, Chapter 3.5, Section 11346.3). When the rulemaking agency receives DOF's letter, it can revise the SRIA to address these comments. The rulemaking agency can file a Notice of Proposed Action with OAL no sooner than 60 days after submitting a SRIA to DOF if the agency previously notified DOF of the intent to promulgate a major regulation by February 1 in the same calendar year.[4] Once the Notice of Proposed Action is published in the California Regulatory Notice Register, the rulemaking agency has one year within which to submit the final rulemaking package to OAL and complete the rulemaking process. During the initial 60-day window after submitting the SRIA to DOF, the rulemaking agency should take note of any omissions or method-

---

[4] If the agency has not notified DOF of the intent to promulgate a major regulation by February 1 of the same calendar year, this window increases from 60 days to 90 days (California Code of Regulations, Title 1, Division 3, Chapter 1, Section 2002[a]).

FIGURE 3

Key Steps in the Rulemaking Process for Major Regulations

Agency identifies the need for regulation, gathers evidence, and consults with experts and the general public as needed.

Agency notifies DOF of intent to promulgate a major regulation (DF-130 form), drafts regulation, and conducts a SRIA.

Agency submits the SRIA and summary (DF-131 form) to DOF.

DOF reviews the SRIA and provides comments to agency (within 30 days).

Agency submits the rulemaking package to OAL.

OAL publishes a Notice of Proposed Action.

Government holds initial public comment period that may include public hearings (45 days).

Agency reviews public comments and revises the regulation and SRIA as appropriate.

Government holds second public comment period if proposed regulation or SRIA has been substantially revised (either 15 or 45 days).

DOF approves fiscal estimates and signs off on proposed regulation.

Agency finalizes the proposed regulation and supporting materials and submits the final rulemaking package to OAL.

OAL conducts final review of the rulemaking package (30 days).

If approved, OAL files notification with the Secretary of State.

The regulation becomes effective on one of four quarterly dates based on the date it is filed, unless otherwise specified.

SOURCE: Features information from OAL, undated-b.

NOTE: Blue indicates a rulemaking agency action, green indicates a DOF action, red indicates an OAL action, orange represents a public comment period, and grey indicates a final regulatory action.

ological issues that would require substantial revisions that might prevent the agency from responding to DOF's comments and finalizing the rulemaking process within one year. In this case, the rulemaking agency has the option to withdraw the SRIA and reinitiate the process at a later time. Doing so might save time and resources compared with failing to complete the rulemaking process within one year and having to restart completely.

OAL assists through the remainder of the rulemaking process. The initial rulemaking package for OAL must include the Notice of Proposed Action, an initial statement of reasons for the proposed regulation, the express terms (i.e., the regulatory text) of the proposed regulation, and the Standard Form 399 ("Economic and Fiscal Impact Statement").[5] In the case of major regulations, this package also includes the SRIA, formal comments from DOF, and agency responses to DOF's comments. Once the Notice of Proposed Action is published in the state register, OAL oversees a 45-day public comment period, which can include a public hearing. Once the initial public comment period ends, the rulemaking agency must respond to comments and revise the regulation and SRIA as appropriate. The summary and response to public comments must be included as part of the agency's Final Statement of Reasons (California Government Code, Chapter 3.5, Section 11346.9). If OAL determines that the changes are substantial, a second comment period begins.

The rulemaking process is completed when DOF has approved the agency's fiscal estimates and signed the Standard Form 399, OAL has reviewed and approved the regulation, and OAL has notified the Secretary of State of the outcome. Unless a different date is provided in statute or state law, the regulation will subsequently become effective on one of four quarterly dates, as shown in Table 2. Rulemaking agencies may also request a later effective date or demonstrate the need for an earlier effective date, subject to approval by OAL.

---

[5] The Standard Form 399 summarizes the agency's calculations of benefits and costs, statewide effects including job impacts, fiscal impacts to state and local governments, and comparison of regulatory alternatives. It also includes the agency's determination of whether the regulatory action is a major regulation. Finance's instructions and code citations for the Standard Form 399 can be found in Sections 6601–6616 of the State Administrative Manual (California Department of General Services, 2014).

TABLE 2

Effective Date of Proposed Major Regulations, According to Date Filed

Effective Date	Date Final Regulation Is Filed with the Secretary of State
January 1	Between September 1 and November 30
April 1	Between December 1 and February 29
July 1	Between March 1 and May 31
October 1	Between June 1 and August 31

SOURCE: Features information from OAL, undated-b.

Although major regulations generally move through this standard process, an accelerated process is available for emergency regulatory actions.[6] The emergency process is intended to be invoked only in cases in which the standard rulemaking process is too slow to accommodate urgently needed actions for public health or safety reasons. The process described previously is generalized for any California executive branch agency. Appendix A provides an overview of the regulatory process for the development of an occupational health standard, which is specific to DIR.

Agencies' Roles in the Rulemaking Process

The administrative rulemaking processes described in this report are unique to California, but the process largely parallels the federal regulatory impact analysis process and might be similar to other states. Twenty-nine state administrative procedure acts require administrative agencies to conduct benefit-cost analyses before implementing regulations (Ballotpedia, undated). In California, the rulemaking agency is primarily responsible for developing the proposed regulation, producing sufficient analytic justifica-

---

[6] If the state agency intends to make an emergency rule permanent and that emergency rule is a major regulation, a SRIA is required. The SRIA must be completed prior to the expiration of the effective period of the emergency regulation to obtain a certificate of compliance rather than allow the emergency regulation to sunset. If a SRIA is needed for the permanent rulemaking, agencies must comply with the standard 60-day (or 90-day) window, depending on the timing of submission of the DF-130; no accelerated timeline is permitted.

tion for any regulation (including an economic analysis), and revising the proposed regulation and any supporting economic analyses as appropriate based on feedback received from DOF, stakeholders, and the general public. DOF generally plays a consulting role by providing methodological advice to ensure that agencies' economic analyses are rigorous and appropriate. Meanwhile, OAL is tasked with steering proposed regulations through the rulemaking process; ensuring that agencies' regulations are clear, necessary, and legally valid; and disseminating the results of the rulemaking process to the general public and the Secretary of State (OAL, undated-a).

Federal agencies have a significantly longer history of conducting regulatory impact analyses. Presidential Executive Order (EO) 12291 (1981) required most executive agencies to perform benefit-cost analyses for major regulations and established a centralized regulatory review process under the Office of Information and Regulatory Affairs within OMB. EO 12291 was revoked in 1993 and replaced with EO 12866 (1993), which remains in effect as of this writing. EO 12866 established more–widely accepted common principles of regulation and streamlined the regulatory process by, for example, eliminating review of nonsignificant rules, clarifying confusing rulemaking issues to resolve disputes and rule delays, and making the process more accessible and open to the public. This led OMB, in cooperation with federal agencies and the public, to develop Circular A-4, the primary regulatory analysis guidance document for federal agencies.[7] Circular A-4 describes the key elements of a regulatory analysis as follows:

- identifying and evaluating the need for the regulatory action;
- defining the baseline;
- identifying a range of regulatory alternatives;
- assessing the benefits and costs of regulatory alternatives by:
  - gathering evidence relevant to the effects of the various alternatives;

---

[7] Circular A-4 was initially issued on September 17, 2003 (OMB, 2003). OMB issued an update to Circular A-4 on November 9, 2023. The effective date of the updated circular is March 1, 2024.

– quantitatively estimating or qualitatively describing the benefits and costs of each regulatory alternative; and
– summarizing the regulatory analysis. (OMB, 2023)

Although this guidance document does not apply to California state agencies, it is a useful resource that describes best practices for benefit-cost analysis and is generally accessible for economists and noneconomists. Some major federal regulatory agencies built on these guidelines, providing a greater level of technical detail and more comprehensive guidance for assessing the impacts of their own rules (U.S. Department of Health and Human Services [HHS], 2016; U.S. Environmental Protection Agency [EPA], 2014).

The Basic Elements of a Standardized Regulatory Impact Assessment

In this section, we provide a high-level overview of the general guidance on how to prepare a SRIA. The SRIA should consider all benefits and costs to help government agencies and the public understand potential trade-offs between policy choices. Although SRIAs necessarily differ in scope and methodology because of the unique aspects of each proposed regulation, they generally contain similar descriptive information and analyses prepared in accordance with California Code of Regulations, Title 1, Section 003. However, state agencies have taken notably different approaches to preparing SRIAs.

DOF's guidelines are generally less prescriptive than guidelines for federal agencies (e.g., DOF does not specify which discount rates should be applied to future benefits and costs). However, the California Code of Regulations imposes several requirements that are specific to evaluating statewide impacts in California and have no counterpart in federal regulatory impact analyses. This report is intended to summarize existing guidance and enhance best practices for preparing California SRIAs. When analytic requirements or methods are not otherwise specified in California guidance and regulations, we draw from best practices for regulatory analysis from OMB and other federal agency sources.

Figure 4 shows the common elements of a SRIA. To use this benefit-cost analysis framework, policymakers should first develop a comprehensive list of potential benefits, costs, and other impacts, then use a screening analysis to determine how to prioritize resources and analyses on the issues that are most important for decisionmaking. More-detailed descriptions of the screening analysis and each subsequent step derived from DOF guidance that is codified in the California Code of Regulations and based on best practices for regulatory analysis are provided in the sections that follow.

FIGURE 4

Basic Elements of a Standardized Regulatory Impact Assessment

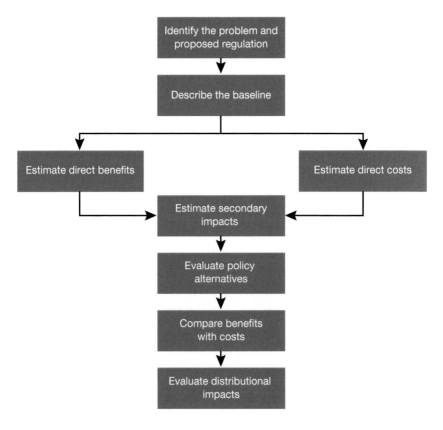

Conduct a Screening Analysis

It could be useful to conduct a screening analysis prior to the development of a SRIA while the proposed regulation is still being conceived. The purpose of a preliminary screening analysis is to inform the development of regulations, such as by identifying and evaluating policy options, and decisions about future research and analysis. HHS states that screening analysis

> is typically based on easily accessible data and simple assumptions; its goal is to provide preliminary information on the possible direction and magnitude of the effects and to inform decisions about future work. For example, high-end values can be used to determine whether various types of outcomes are likely to be significant even under extreme assumptions. Depending on the results, this screening may be followed by more detailed assessment that involves collecting additional data, refining the methods used, and possibly expanding the scope of the analysis . . . . (HHS, 2016, p. 9)

In the context of California regulations, a screening analysis could help determine whether a formal SRIA is needed, what type of macroeconomic model (if any) is most appropriate for the analysis, and what additional research and analysis might be needed to reduce uncertainty. Screening analysis can help inform decisions about how to prioritize limited resources so that efforts to conduct more-detailed analyses, monetize hard-to-quantify benefits, or acquire additional data to evaluate distributional impacts (e.g., concerns about social equity) are proportionate to their relative importance to decisionmaking.

Identify the Problem and Proposed Regulation

DOF requires that agencies provide a statement of the need for the proposed major regulation, summarize who will be affected by the proposed regulation, and summarize these impacts in economic terms (California Code of Regulations, Title 1, Division 3, Chapter 1, Section 2002[c]). The California Code of Regulations does not specify how the need for the regulation should be demonstrated or justified. However, federal guidance for regulatory analyses suggests several approaches, such as explaining whether the proposed regulation is designed to correct a market failure, address behavioral biases, improve government operations and service delivery, or promote distributional fairness and advance equity (OMB, 2023).[8]

Regulatory analyses are intended to provide agencies and the public with information to determine whether a proposed regulation is an "efficient and effective means of implementing the policy decisions enacted in statute or by other provisions of law in the least burdensome manner" (California Government Code, Chapter 3.5, Section 11346.3). Therefore, a SRIA should identify the specific authority for the proposed regulatory action and the intended benefits included in the authorizing statute.[9] Evidence, preferably peer-reviewed research, should be cited to support the likely causal link between the proposed regulatory requirements and their desired outcomes (California Code of Regulations, Title 1, Division 3, Chapter 1, Section 2003[e] [5]; HHS, 2016). Federal guidance states that when a rule includes several distinct provisions, agencies should examine how each provision modifies existing requirements and, to the extent feasible and appropriate, analyze the benefits and costs of each provision separately (OMB, 2023).

Preliminary analyses conducted during the development phase of a regulation can help agencies identify the policy options that are most likely to achieve desired policy outcomes and maximize net benefits to society. The SRIA should document how those and subsequent analyses contributed to the development of the proposed regulation, such as by evaluating the impacts of a variety of policy options or regulatory tools that are available to achieve policy objectives. That is, SRIAs should not be conducted only after the fact, once the proposed regulation is selected.

Describe the Baseline

DOF directs agencies to identify an established baseline "that reflects the anticipated behavior of individuals and businesses in the absence of the proposed major regulation" (California Code of Regulations, Title 1, Division 3, Chapter 1, Section 2003[d]). The SRIA should identify the expected start

---

[8] Standard Form 399, which is required for all regular and emergency regulations, similarly requires a statement of need for the proposed major regulation.

[9] As noted previously, the analysis should address all costs and benefits, not only those that are included in the statute.

date for the proposed regulation and the agency's estimate of when the regulation will be fully implemented. A key requirement of benefit-cost analysis is that each policy option should be compared against a common *no regulatory action* baseline (i.e., the status quo).[10] As a general rule, the baseline should do the following:

1. **Characterize the universe of stakeholders (i.e., individuals, business enterprises, and state and local government agencies that will be affected by the proposed regulation).** When there are anticipated uneven distributional consequences of the proposed regulation, the analysis should endeavor to report how the impacts of the regulation are likely to differ among identifiable groups. The baseline should provide information on the relevant characteristics of those individuals (e.g., income, occupation, geography, race, sex, age) and/or businesses (e.g., industry, business size, location, labor demographics). The analysis should include estimates of the number of stakeholders affected across each group.

2. **Reflect anticipated industry or demographic trends without the regulatory action.** The baseline should not assume that existing conditions will persist in the future. Instead, it should reflect anticipated economic and/or demographic conditions in the absence of the proposed regulation (e.g., changes in the size and/or composition of the affected population, the economy, technology available).

3. **Address areas of overlap or conflict with other laws and regulations.** The baseline must discuss the effects of any anticipated interaction with other laws, regulations, or policies, including those that have been enacted but have not yet been implemented. Specifically, analysts must determine how state or federal regulations will affect market conditions that directly influence the benefits or costs associated with the proposed regulation.

4. **Clearly state assumptions regarding compliance rates with existing regulations**. A practical approach might be to assume full compliance with existing and recently enacted regulations so as to

---

[10] The status quo should not be considered a separate regulatory alternative; all policy options must be compared against the same no-regulatory-action baseline.

focus on the incremental impacts of the proposed regulation without potentially double-counting the benefits or costs captured by analyses conducted for other rules (EPA, 2014). However, when the degree and extent of noncompliance can affect the evaluation of policy options and there are sufficient monitoring data available, it might be useful to use a partial-compliance baseline that reflects available information on compliance rates. For example, a proposed regulation might amend or seek to address suspected noncompliance with an existing regulation. Analysts might have to assess the degree of noncompliance and whether compliance rates vary systematically to develop a reasonable baseline. When more than one baseline is reasonable, multiple baselines can be used; however, caution should be taken to clearly communicate the results of the analyses because the number of scenarios and comparisons of regulatory alternatives will multiply.

Describe the Economic Methodology and Approach

Benefit-cost analysis provides a standard framework with which to assess the potential impacts of a proposed regulation. The scope and scale of those impacts might inform decisions about which economic methodologies to employ to evaluate them. The State Administrative Manual (Sections 6600–6616) contains regulations that pertain to conducting SRIAs, including estimating costs. DOF directs agencies to both qualitatively and quantitatively describe estimated changes in the behavior of businesses and/or individuals in response to the proposed regulation and, if feasible, the extent to which benefits or costs are passed onto others (California Code of Regulations, Title 1, Section 2003[f]). The analysis must provide estimates of the number of individuals and entities who would be affected each year and information on their characteristics to assess disparate impacts.

OMB (2023) provides specific guidance to federal agencies on minimum quality standards for regulatory analysis, which are also useful for state agencies to emulate, such as the following:

- clearly state "the key methods, data, and other analytical choices you make in your analysis" (OMB, 2023).

- provide "documentation that the analysis reflects the highest quality evidence (including scientific, technical, economic, and indigenous knowledge) and analytic methods" (OMB, 2023).
- "seek out the opinions of those who will be affected by the regulation as well as the views of those individuals and organizations who may not be affected but have special knowledge or insight into the regulatory issues" (OMB, 2023).
- address uncertainties associated with the estimates if quantified benefit or cost estimates depend heavily on certain assumptions. Options for addressing uncertainty in increasing levels of complexity include (1) providing a qualitative discussion of the likely direction and magnitude of potential biases; (2) conducting a numerical sensitivity analysis to examine how the results of the analysis vary with plausible changes in assumptions, data inputs, and alternative analytical approaches; and (3) conducting a probabilistic analysis using simulation models (e.g., Monte Carlo simulation) to quantify the statistical distribution of economic impacts.
- use nonmarket valuation methods (e.g., stated or revealed preference studies) to measure impacts that cannot be fully monetized using standard market prices, including reductions in the risk of fatal and non-fatal injuries or illnesses.

The selected economic approach should be used to estimate benefits and costs over a multi-year time period and must include impacts that accrue in first 12 months of the regulation through at least 12 months after the regulation is anticipated to be fully implemented (California Code of Regulations, Title 1, Section 2002). A SRIA must also assess impacts to the state and affected local government agencies that are attributable to the proposed regulation, such as the cost of enforcement and compliance and any government savings (California Department of General Services, 2014, Section 6607). Furthermore, DOF directs agencies to address how the impacts of the regulation will be distributed over time if there are significant differences in the timing of benefits and costs (California Code of Regulations, Title 1, Section 2003[e][4]). Although there is no formal guidance around selecting an appropriate analytical time horizon, the analysis should ideally cover a period that is long enough to capture all of the important costs and benefits of the proposed regulation, particularly in cases in which there could be significant up-front costs but the benefits might not be realized for years.

In some cases, those subject to the regulation might comply before the regulation becomes effective. The analysis should include the impacts of early compliance in anticipation of the regulation and document the extent to which this precompliance is consistent with the regulation's final provisions; for example, it might exceed the ultimate requirements. Agencies might find it useful to disaggregate the results and report effects of such early compliance separately from the impacts of the regulation. However, long-standing practices used by any segment of the population subject to the rule that are already compatible with the proposed regulation might be accounted for in the baseline and not attributed to the regulation.

The appropriate approach for estimating the monetary value of both benefits and costs is based on the concept of opportunity cost (HHS, 2016; OMB, 2023). The use of any resource (e.g., capital, labor) has an opportunity cost that is equivalent to the value of the best use of that resource. These values are generally based on estimates of individuals' willingness to exchange their income for effects that they themselves experience, expressed as willingness-to-pay or willingness-to-accept compensation. Although theory suggests that estimates of these two measures should be similar in most cases, empirical work sometimes finds large differences that are not well-understood. Methods for nonmarket valuation are discussed in more detail in later sections, as well as in EPA (2014), HHS (2016), and Robinson et al. (2019).

DOF directs agencies to attempt to quantify and monetize all benefits and costs but to include a qualitative description whenever quantification is impossible or impractical. However, as a practical consideration, DOF has generally interpreted State Administrative Manual Section 6607 to require that all costs be quantified and any assumptions clearly stated if reliable cost data are not readily obtainable. Benefits, however, must be quantified only to the extent feasible. Sufficiently addressing these impacts might have important consequences for decisionmaking.

The California legislature directed DOF to instruct state agencies to, at a minimum, assess the value of nonmonetary benefits "such as the protection of public health and safety, worker safety, or the environment, the prevention of discrimination, the promotion of fairness or social equity . . . and other nonmonetary benefits" (California Government Code, Chapter 3.5, Section 11346.36). DOF directed agencies to quantity these benefits to the extent feasible (California Code of Regulations, Title 1, Section 2003[a][3]). Therefore, it is necessary to identify unquantified impacts and discuss them qualitatively as with other impacts.

However, federal guidance and alternative sources are available to quantify and monetize some of these benefits, such as improvements in worker safety and environmental protection. A purely qualitative analysis could lead decisionmakers to rely on subjective judgment to weigh unquantified impacts against other impacts. Where feasible, some degree of quantification of impacts—which are generally measured in physical units (e.g., number of exposures to known hazards)—might help signal the relative importance of different types of impacts even if such impacts cannot be monetized (HHS, 2016). In other cases, measuring intangible impacts of the regulation (e.g., dignity, employee morale, equity) might not be feasible, but it could be possible to count the number of individuals or businesses affected or report other intermediate measures.

Potential options for incorporating unquantified impacts in a SRIA will depend on the available data and might include the following methods (OMB, 2023; HHS, 2016):

- **Qualitative discussion:** Any approach (including those described in this list) should include a description and discussion of unquantified impacts. To the extent possible, the approach should summarize available information on the likely direction (e.g., positive or negative), magnitude (e.g., high or low), and incidence (e.g., highly probable or improbable) of those impacts; the strength of evidence for causal links between the potential impacts and the proposed regulation; and other attributes that are relevant to decisionmaking.
- **Breakeven analysis:** *Breakeven* or *threshold* analysis answers the question "how large would the unquantified benefits need to be for the benefits of a regulation to equal or exceed its costs?" The breakeven threshold, which can only be calculated for a single outcome of the regulation at a time, can provide useful information for decisionmaking when data are available on the consequence of the impact (i.e., monetized

value) but not the frequency (i.e., number of cases averted). Breakeven analysis relies on subjective judgment but provides a basis—which is informed by the potential magnitude of the impact—for determining whether nonmonetized impacts could plausibly exceed the breakeven threshold.

- **Cost-effective analysis:** A *cost-effective* analysis compares the relative costs of policy options that have the same intended outcome when all the relevant benefits cannot be monetized. This approach might be useful for providing insights when an impact can be quantified in physical units (e.g., number of cases averted) but not easily expressed in terms of dollars. However, cost-effective analysis has limited value when multiple types of impacts must be considered, and it cannot determine whether the benefits of a regulation exceed its costs or which policy option is likely to maximize the net benefits to society.

- **Bounding analysis:** A *bounding* or *what-if* analysis compares the impact of hypothetical but plausible scenarios with the net benefits of the proposed regulation using lower- and upper-bound estimates for the magnitude of the unquantified impacts. Although it is similar to a numerical sensitivity analysis, a bounding analysis might present a wide variety of estimates using relatively little supporting evidence and should be presented separately from the primary estimates because of the high degree of speculation involved.

Estimate Direct Costs

Costs represent the total burden a regulation will have on the economy—that is, the sum of all opportunity costs because of the regulation. For example, a new administrative requirement might require one hour of a manager's time that could otherwise be focused on maximizing production or operational efficiencies. These hours should be aggregated across employees and businesses then monetized using a standard methodology that uses statewide occupation-based wage and salary information, such as the information reported by the U.S. Bureau of Labor Statistics (BLS).[11] For state employee wages, analysts can use the California Department of Human Resources' pay scale (California Department of Human Resources, 2020).

---

[11] See, for example, BLS, undated; and BLS, 2023.

DOF does not provide specific guidance on estimating the monetary value of time. Several federal agencies offer guidance on valuing time in regulatory impact analyses (Baxter, Robinson, and Hammitt, 2017; EPA, 2020; U.S. Department of Transportation, 2016). These approaches differ in certain respects; however, resolving those differences is beyond the scope of this report. These differences should be reviewed by analysts and addressed through sensitivity analysis if they significantly affect the results of the benefit-cost analysis. Generally, labor costs (e.g., employees undertaking compliance activities) should include hourly wages, fringe benefits (including voluntary and legally required benefits), and indirect or overhead costs.[12] Regulations that impose an administrative burden on individuals without compensation are generally estimated using the post-tax wages that an individual would have been compensated for working (and, in some cases, the voluntary portion of fringe benefits received).

Analysts should also estimate costs and cost savings associated with changes in behavior by businesses and/or individuals in response to the proposed regulation, including the extent to which costs are passed onto others, if feasible (California Code of Regulations, Title 1, Section 2003[f]). Separate cost and savings calculations, as appropriate, must be provided for federal, state, and local government agencies (California Department of General Services, 2014). As a general practice, cost calculations should include both the up-front, one-time costs (e.g., capital expenditures) and recurring costs (e.g., operations and maintenance, annual training) of complying with the proposed regulation. Capital expenditures and other direct costs that are attributable to the proposed regulation should be fully accounted for using the best available estimates from peer-reviewed literature or indus-

---

[12] Benefits include paid time off, health insurance, retirement contributions, and payroll taxes (i.e., Social Security, Medicare, unemployment insurance, workers' compensation). Benefit estimates can be obtained from the BLS' "Employer Costs for Employee Compensation" (undated). In this context, indirect labor costs generally include administrative costs (e.g., human resources functions), office and equipment expenses, and employee development activities. The U.S. Department of Labor offers guidance on estimating overhead costs by using industry-wide data on costs obtained from the U.S. Census Bureau's Annual Survey of Manufacturers and Service Annual Survey (U.S. Department of Labor, 2016). Data for state and local government agencies are generally unavailable.

try sources, such as a manufacturer's suggested retail price, because these costs could limit financial resources that would otherwise be available for investment in other productive areas. Analysts should also consider costs and cost savings associated with behavioral responses to the implications of the proposed regulation, such as health risks or risk reductions that might be associated with changes in production methods or a transition to alternative products.

DOF requires that a SRIA includes estimates of the proposed regulation's fiscal impacts on state and local government agencies, including effects on local governments and the state's General Fund and special funds (California Code of Regulations, Title 1, Section 2003[h]). These estimates must include the costs associated with compliance, administration, implementation, and enforcement; any savings; and any other impacts, such as changes in sales tax revenues (California Department of General Services, 2014). There is little additional state guidance around approaches for estimating fiscal impacts to state and local governments. Furthermore, standard tools for economic impact analysis generally use less transparent methods and provide minimal (in some cases zero) information for estimating tax revenue impacts. Agencies have used a variety of approaches to estimate fiscal impacts, but the failure to include all fiscal impacts to state and local government agencies is the one of DOF's most frequent critiques of SRIAs (see discussion in the "Department of Finance Comments on Past Standardized Regulatory Impact Assessments" section).

Estimate Direct Benefits

Generally, impacts related to policy outcomes (e.g., reductions in health risks, improvements in environmental quality, improvements in public safety, increased access to services) should be categorized as benefits. California Government Code, Chapter 3.5, Section 11346.3(c)(1)(F) requires state agencies to prepare SRIAs that evaluate "the benefits of the regulation, including, but not limited to, benefits to the health, safety, and welfare of California residents, worker safety, and the state's environment and quality of life, among any other benefits identified by the agency." Estimating the benefits of proposed regulations in monetary terms facilitates comparison of the types of benefits in the same units of measurement and allows the calculation of net benefits so that policy options can be compared with the baseline and one another. However, in many cases, benefits cannot be monetized using market prices. There are alternative methods to monetize certain benefits and ways to incorporate nonmonetized impacts into decision-making (discussed previously in the "Describe the Economic Methodology and Approach" section).

Revealed- or stated-preference methods are generally used to estimate the monetary value of health-related impacts, including avoided injuries, illnesses, and fatalities (EPA, 2014; HHS, 2016; Robinson et al., 2019). *Revealed-preference methods* estimate the value of nonmarket outcomes (e.g., fatal or nonfatal injury risk reduction) using consumer purchase decisions or employment decisions at prevailing market prices. Because the risk reduction is not purchased directly, the value must be inferred from market decisions in which safety is a factor in an individual's decision. For example, hedonic wage studies examine compensation differentials between jobs that involve different levels of risk of injury or death using statistical methods to isolate the effects of those risks from other factors (HHS, 2016). *Stated-preference methods* use surveys to estimate values using subjects' responses to hypothetical scenarios.

Three major federal regulatory agencies provide guidance on how to value impacts on health and longevity using revealed- and stated-preference studies: the EPA, HHS, and the Department of Transportation (EPA, 2014; HHS, 2016; HHS, 2021; U.S. Department of Transportation, 2021). When conducting SRIAs, analysts generally rely on this federal guidance to ensure consistent and justifiable approaches across regulations. The HHS and EPA guidance documents and Robinson et al. (2019) provide information on the use of market measures (e.g., estimates of the direct and indirect costs of illness) as proxies in cases in which more appropriate and comprehensive nonmarket valuation measures are not available. In addition, the U.S. government has developed estimates for valuing reductions in greenhouse gas emissions (Interagency Working Group on the Social Cost of Carbon, 2021) that are used in regulatory analyses across the federal government. In November 2023, the EPA released a new set of estimates for the social cost of greenhouse gas emissions incorporating comments from an external peer review panel assessing the agency's 2022 draft updated estimates (EPA,

2023). The Organisation for Economic Co-operation and Development also recently developed new estimates for several health conditions.[13]

Benefits that accrue to all affected stakeholders should be described quantitatively (to the extent possible) and qualitatively, including for state and local government agencies in the analysis of fiscal impacts.

Estimate Secondary Impacts

DOF requires agencies to use an economic impact method that can generate California-specific estimates of employment by economic sector and other measures of state economic output using a set of inter-industry relationships (California Code of Regulations, Title 1, Section 2003[a]).[14] DOF defines *economic impact* as "all costs or all benefits (direct, indirect and induced) of the proposed major regulation on business enterprises and individuals located in or doing business in California" (California Code of Regulations, Title 1, Section 2000). Although these terms are not consistently described in the economic literature, in this context these concepts are typically defined as follows:

- *Direct effects* are incremental changes made by businesses or individuals in response to a regulatory action in production or expenditures on goods and services provided by other businesses. These changes can be positive or negative.
- *Indirect effects* are impacts from business-to-business purchases in the local supply chain (across all interconnected industries) stemming from changes in the initial industry's spending with their suppliers.
- *Induced effects* are impacts associated with household spending as a result of changes in labor income (i.e., wages after taxes, savings, commuter income) caused by direct and indirect effects.

---

[13] The Organisation for Economic Co-operation and Development has published estimates and methodology reports to support socioeconomic analyses of the regulation of chemicals by helping to better quantify and monetize morbidity because of and environmental impacts from exposure to chemicals (Organisation for Economic Co-operation and Development, undated).

[14] A detailed review of economic approaches is provided in the "Review and Selection of Macroeconomic Models" section.

The economic methodology generally follows a standard approach to separately reporting direct, indirect, and induced impacts of a proposed regulation. DOF requires that SRIAs

> produce (to the extent possible) quantitative estimates of economic variables that address or facilitate the quantitative or qualitative estimation of the following:
>
> (A) The creation or elimination of jobs within the state;
>
> (B) The creation of new businesses or the elimination of existing businesses within the state;
>
> (C) The competitive advantages or disadvantages for businesses currently doing business within the state;
>
> (D) The increase or decrease of investment in the state;
>
> (E) The incentives for innovation in products, materials, or processes; and,
>
> (F) The benefits of the regulations, including, but not limited to, benefits to the health, safety, and welfare of California residents, worker safety, and the state's environment and quality of life, among any other benefits identified by the agency. (California Code of Regulations, Title 1, Section 2003[a][3])

Analysts must also provide information on the model used, inputs, key assumptions, and limitations of the methodology. Several California state agencies outsource this analysis to consultants who specialize in using existing economic tools to evaluate statewide policy impacts or who have developed their own macroeconomic models for policy analysis. Regardless of the economic model selected, quantitative estimates often cannot be reliably produced, and a qualitative discussion of potential impacts is necessary. Any qualitative discussion should describe the likely direction and magnitude of the economic impact and disclose key assumptions or uncertainties that might be useful for decisionmakers to consider.

Evaluate Policy Alternatives

Analysts must identify and evaluate a variety of policy options or regulatory tools available to the agency to achieve the objectives of the proposed regulation. DOF directs agencies to evaluate a preferred approach and at least two regulatory alternatives, when possible, including

- An alternative that could achieve additional benefits beyond those associated with the proposed major regulation; and
- A next-best alternative that would not yield the same level of benefits associated with the proposed major regulation, or is less likely to yield the same level of benefits. (California Code of Regulations, Title 1, Section 2003[e][2])

However, if there are no feasible and reasonable alternatives allowable under statute, "an agency is not required to artificially construct alternatives or describe unreasonable alternatives" (California Government Code, Chapter 3.5, Section 11346.2). The no-regulatory-action scenario should not be considered as one of the two regulatory alternatives because it is the baseline against which all other policy options must be compared. A wider variety of policies may be evaluated, although it is unnecessary to compare policies that are nearly identical. The SRIA must describe and evaluate each regulatory alternative to the proposed regulation and state the agency's rationale for rejecting those alternatives (California Code of Regulations, Title 1, Section 2003[e][1]).

DOF directs agencies to consider a comparison of the cost-effectiveness of alternatives (California Code of Regulations, Title 1, Section 2003[e][3]). The economic analysis should inform agencies and the public whether regulatory choices are "an efficient and effective means" of achieving policy objectives in the least burdensome manner (California Government Code, Chapter 3.5, Section 11346.3). However, in certain cases, having a significant adverse economic impact or negative net benefits would not necessarily invalidate a proposed regulatory action. A frequent shortcoming of SRIAs is that agencies either fail to identify or evaluate a sufficient number of reasonable alternatives, which makes it challenging to determine that the preferred regulatory alternative achieves the greatest net benefits or produces the most desirable outcomes (we discuss this shortcoming further in the "Department of Finance Comments on Past Standardized Regulatory Impact Assessments" section).

Compare Benefits and Costs

Generally, the final step in developing a SRIA is to compare total benefits and total costs. Benefits and costs that cannot be reliably monetized—for example, prevention of discrimination, promotion of distributional fairness or social equity, or an increase in the openness and transparency of business and government—should also be considered. Estimates of benefits and costs should be reported separately for businesses, individuals, and state and local government agencies on an annual basis (i.e., capital costs should not be annualized across multiple years). The annual stream of costs and benefits should be reported over the entire period of analysis, which must extend for at least 12 months after the proposed regulation is estimated to be fully implemented.

The standard methodology to account for differences in the timing of benefits and costs is to discount future impacts to reflect time preferences and opportunity costs of investments made in different periods (EPA, 2014; HHS, 2016; OMB, 2023). Discounting reflects individuals' general preferences to receive benefits sooner and to bear costs later in time and recognizes that costs incurred today are more expensive than future costs because businesses must forgo an expected rate of return on investment of that capital (OMB, 2023). For direct comparison, benefits and costs should be discounted at the same rate. DOF does not provide specific guidance on the choice of discount rates in SRIAs. However, in practice, DOF asks state agencies to justify why the selected discount rate is appropriate and consider using a sensitivity analysis when assumptions regarding the discount rate might materially change findings in the economic analysis.

Additional guidance on the use of discount rates in SRIAs would help ensure a consistent and justifiable approach across regulations. This guidance does not necessarily need to recommend a single value but could instead refer to federal guidance and other recommended best practices. OMB previously recommended using discount rates of 3 percent and 7 percent for regulatory analysis (OMB, 2003). The 7-percent discount rate was intended to reflect the average pre-tax rate of return on private capital in the U.S. economy as estimated in 1992. The 3-percent discount rate was intended to represent the *social rate of time preference*—the rate at which society discounts future consumption—which OMB approximates using the real rate of return on long-term government debt. As of November 2023, OMB recommends using a single discount rate of 2 percent as a base case (OMB, 2023).[15] OMB intends to publish updates to this estimate every three years. The National Academies of Sciences, Engineering, and Medicine (Steurele and Jackson, 2016) recommends using a social discount rate of 3 percent either as a base case or in a sensitivity analysis for economic evaluations. Declining discount rates for longer-term impacts are often recommended in these and other guidance documents. Existing best practices suggest using a reference case of between about 2 percent and 7 percent while examining the sensitivity of the results of the benefit-cost analysis to alternative discount rates through a numeric sensitivity analysis or Monte Carlo simulation. Analysis of regulations that are likely to have intergenerational impacts should allow for lower discount rates.

The results of the economic analysis are generally reported in terms of net benefits. These results can also be reported as a benefit-cost ratio or rate of return, although these approaches generally less preferable. These approaches might be misleading because ratios do not indicate the magnitude of the impact of a regulation and are highly sensitive to whether cost savings or countervailing risks are included in the costs or benefits of the regulation. Ratios can also be misleading when comparing two or more alternatives that have different costs; the net benefits of a higher-cost regulation might be greater than a lower-cost alternative, while the benefit-cost ratio might not be greater. Such a comparison could lead to inappropriate selection of a less cost-effective policy alternative. Finally, using supporting evidence, the analysis should demonstrate that no feasible and reasonable regulatory alternative allowable under statute would be less burdensome

---

[15] This value is an updated estimate of the social rate of time preference, which OMB calculated using (1) the 30-year average of the yield on 10-year Treasury marketable securities—this rate averaged 1.7 percent per year in real terms on a pre-tax basis between 1993 and 2022—and (2) an inflation factor, as measured by the personal consumption expenditure inflation index, of approximately 0.3 percent per year.

and equally effective in achieving the objectives of the authorizing statue or other law being implemented or made specific by the proposed regulation.

Evaluate Distributional Impacts

Distributional analysis helps identify trade-offs between economic efficiency (i.e., maximizing net benefits) and distributional concerns. In some cases, agencies might choose a less efficient policy option to mitigate distributional impacts or achieve other policy objectives (e.g., improving fairness or social equity). To the extent that the impacts of a regulation are significantly different across groups, DOF requires SRIAs to address distributional effects. Specifically, DOFs states that "[c]osts and benefits shall be separately identified for different groups of agencies, businesses and individuals if the impact of the regulation will differ significantly among identifiable groups" (California Code of Regulations, Title 1, Section 2003[c]). DOF further instructs that

> If there are significant differences between the incidence or timing of costs and benefits of a regulation, distributional effects should be addressed, including how the effects of the regulation are distributed, for example, by industry, income, race, sex, or geography, and how the effects are distributed over time. (California Code of Regulations, Title 1, Section 2003[e][4])

In assessing impacts to California businesses, a SRIA must report the distribution of impacts across industries (e.g., by North American Industry Classification System [NAICS] code), as well as by business size. Specifically, a SRIA must identify the number of small businesses affected and the typical compliance costs for a small business to determine whether the proposed regulation will have a significant economic impact on a substantial number of small entities.[16] In California, for the purposes of conducting

---

[16] This is similar to the regulatory analysis requirements for federal agencies under the Regulatory Flexibility Act (Pub. L. 96-354, 1980), amended by the Small Business Regulatory Enforcement Fairness Act (Pub. L. 104-121, 1996), the Dodd-Frank Wall Street Reform and Consumer Protection Act (Pub. L. 111-203, 2010), and the Small Business Jobs Act of 2010 (Pub. L. 111-240, 2010).

an economic impact assessment, a *small business* is a business that (1) is independently owned and operated, (2) is not dominant in its field of operation, and (3) has fewer than 100 employees (California Government Code, Chapter 3.5, Section 11346.3). If relevant to the rulemaking, disparate business impacts across geographic areas should be considered. Agencies are also required to estimate fiscal impacts of the regulation, providing separate calculations for state and local agencies, including compliance and/or enforcement costs and any savings to each level of government (California General Services Department, 2014).

In assessing impacts to individuals, a SRIA should report benefits and costs on a disaggregated basis across demographic groups. For example, analysts can use tables describing the total and percentage of benefits, costs, and net benefits that accrue to different groups (e.g., income percentile, race, sex, geography). To obtain more detailed demographic data, agencies might rely on the Current Population Survey, a comprehensive survey of households sponsored by the Census Bureau for the Bureau of Labor Statistics that provides information on the labor force, employment, unemployment, earnings, and other demographic and labor force characteristics (U.S. Census Bureau, 2023).

Similarly, EO 12866, as amended by EO 14094, directs regulatory analyses of major federal rulemakings to recognize distributive impacts and equity, to the extent permitted by law. Federal guidance defines *distributional effects* as "how the benefits and costs of regulatory action are ultimately experienced across the population and economy, divided up in various ways (e.g., income groups, race or ethnicity, sex, gender, sexual orientation, disability, occupation, or geography; or relevant categories for firms, including firm size and industrial sector)" (OMB, 2023) Additional guidance is provided in EPA (2014), HHS (2016), and Robinson et al. (2019). Although EO 12866 and 13563 noted that agencies may include distributional analysis in their regulatory analyses, a review of past regulatory analyses that was focused on regulations that lead to health-risk reductions found that federal agencies generally provided little information on distributional impacts. The review attributed the lack of analysis of the distribution of benefits and costs to a philosophically motivated focus on efficiency (i.e., maximizing net benefits), concerns about political or legal implications, an assumption that distributional impacts are small or insignificant, or data and resource constraints (Robinson, Hammitt, and Zeckhauser, 2016). The update to Circular A-4 provides further guidance and options for including distributional analysis in federal regulatory analyses (OMB, 2023).

Address Uncertainty

At each stage in the process, analysts should address key sources of uncertainty qualitatively and quantitatively (EPA, 2014; HHS, 2016; OMB, 2023; Robinson et al., 2019). Decisionmakers should be able to understand the primary sources of uncertainty in the analysis and how those sources of uncertainty affect the results. The models and approaches used to estimate benefits and costs might be limited by the accessibility and quality of data; the extent to which the data address the same population, industries, or geographic area as the study population; the availability of relevant scientific research; and uncertainty regarding future projections of economic and other conditions. As a best practice, analysts should address the relative cost-effectiveness of policy alternatives and the extent to which uncertainty around specific parameters affects the likelihood that a particular policy will yield benefits that exceed costs. Nonmonetized benefits should also be discussed, indicating the likely direction (e.g., positive or negative) and magnitude (e.g., large or small) of the impact. Because it might be infeasible or impractical to evaluate alternative values for all of the assumptions in the analysis, analysts should focus on the assumptions that have the largest impact on the final results and the most significant implications for decisionmaking.

There are several options—which vary in terms of complexity—for addressing uncertainty, as discussed in the sources referenced previously in this section and explored in more detail in HHS (2016). For SRIAs of California regulations, three approaches should be considered. First, when additional research is not feasible because of time or resource constraints, a SRIA should qualitatively discuss major sources of uncertainty and their implications. With regard to costs, DOF has generally directed California state agencies to make and clearly state assumptions when market data are not readily obtainable. In the case of nonmonetized benefits, qualitative discussions are generally permitted. In any qualitative discussion, analysts should disclose potential biases (i.e., under- or over-estimation) in the assumptions, including the direction of the bias and the likely magnitude of its effect.

Second, when potential biases can be characterized numerically, analysts should consider a quantitative sensitivity analysis and vary one or several parameter values to calculate alternative results for direct comparison. A sensitivity analysis should generally include (1) a central tendency estimate (e.g., a mean or median) or best estimate and (2) a range or confidence interval. A quantitative analysis ideally describes the probabilities of relevant outcomes and estimates of the monetized consequences of the projected outcomes. If there is a lack of evidence or scientific consensus, the analysis might be able to only describe the benefits and costs under plausible scenarios and characterize the assumptions underlying each alternative scenario without assessing the relatively likelihood of each outcome. In cases in which more than one baseline is reasonable and the underlying assumptions will significantly affect the estimated benefits and costs, analysts should consider using multiple baselines to demonstrate what the world might look like under various scenarios. However, in all cases, benefits and costs should be evaluated against the same baseline (California Code of Regulations, Title 1, Section 2002[d]; OMB, 2023).

Finally, more-sophisticated probabilistic analysis that relies on statistical distributions of parameter values (e.g., Monte Carlo simulation) could be useful in cases in which there are multiple, large uncertainties that present challenges for estimating direct impacts or for cases in which there are cascading effects. Approaches to evaluating the distribution of results are feasible when more information is available on the distribution of uncertain parameters and any dependencies among them. Expert elicitation could be a useful approach for quantifying the effects of uncertainty or developing probability distributions if data are limited or there are gaps in the scientific literature (Cooke, 1991; Morgan and Henrion, 1990; O'Hagan et al., 2006).

Review and Selection of Macroeconomic Models for Standardized Regulatory Impact Assessments

DOF requires that agencies use an economic approach that can estimate statewide regulatory impacts on economic outputs (such as personal income, jobs by economic sector, exports and imports, and gross state product) "based on inter-industry relationships that are equivalent in structure to the Regional Input-Output Modeling System [RIMS II] published by the Bureau of Economic Analysis" (California Code of Regulations, Title 1, Section 2003[a][2]). This requirement directs agencies to use one tool from a limited number of commercially available economic modeling tools to evaluate how regulatory actions will affect the state economy. These tools vary in terms of cost, complexity, transparency, adaptability, level of aggregation (i.e., industry detail), and forecasting capabilities. All the models generally require some level of expertise or substantial training to use. The economic models generally fall into two classes: input-output (I-O) models or computable general equilibrium (CGE) models.

Regional or national I-O tables represent links between sectors of the economy;[17] these tables capture the goods and services produced by each industry and the use of these goods and services by other industries and final users, including households and government. Using data from the BLS and the U.S. Bureau of Economic Analysis (BEA), I-O models provide a set of multipliers to create a static representation of a region's economy at given a point in time for use in economic impact analyses to estimate the total impact of a policy change or other activity in a region. Economic impacts can be expressed in terms of total output, value added (gross domestic product), earnings, or employment for individual industries and across the entire regional economy. Regional purchase coefficients in an I-O model describe the extent to which industries in an area of interest (e.g., state, region, country) use inputs from within the local economy versus outside it (i.e., imports used to meet demand). I-O models can be used to compare the impact of regulatory alternatives on a single industry or many industries to understand which policies result in the greater spillovers to the regional economy or have proportionally larger effects on employment.

CGE models are more complex and statistically rigorous tools. These models can link an I-O model to an econometric or structural forecasting model to account for dynamic changes in the regional economy over time.

---

[17] I-O models are based on methods developed by Wassily Leontief in the late 1930s, for which he received the Sveriges Riksbank Prize in Economic Sciences in Memory of Alfred Nobel (commonly referred to as the Nobel Prize in Economics) in 1973.

They contain a set of equations that use economic theory to capture welfare and profit maximization behavior by households and businesses. CGE models allow for greater substitution using prices than I-O models allow. I-O models assume fixed sets of inputs and outputs; CGE models simulate the interdependent relationships between industries by tracking the flow of all money and commodities across the economy, accounting for the interactions between households and industries. However, CGE models do so at an aggregate level and generally provide less granular detail than I-O models. CGE models were developed to account for major economic shifts in the structure of the economy and track dynamic changes in wages and prices over time.

CGE models require additional assumptions, equations, and programming, and, because of their complexity, they are less transparent than I-O models. As a result, successful use of CGE models commonly requires additional specialized training. However, CGE models are more flexible than I-O models; CGE models allow for the analysis of nonmarginal changes to an economy, but I-O models cannot capture substitutions that might result in changes in capital investments or labor. As a result of these relative strengths and weaknesses, I-O models work well for marginal changes in localized areas, while CGE models are best used for large scale changes that occur over time.

The three most widely used, commercially available economic models for economic impact analysis are

- RIMS II
- Impact Analysis for Planning (IMPLAN)
- Regional Economic Models, Inc. (REMI).

Individual datasets (from the local, state, or national level) and multipliers must be purchased on an annual basis to provide up-to-date information. DOF has a license for RIMS II multipliers from BEA that can be provided to state agencies on request. Trade-offs between each of the tools should be carefully considered by agencies that conduct economic impact analyses.

RIMS II

The RIMS II model provides a set of tables of multipliers developed by the BEA. These multipliers can be used to estimate the total regional economic impact of a policy change. It is the simplest to use and the least expensive model. RIMS II contains Type I multipliers—which account for the direct (initial) and indirect (subsequent inputs purchased by supporting industries) spending because of a change in final demand—and Type II multipliers—which account for inter-industry effects and induced impact related to the spending of workers whose earnings are affected by a change in final demand. RIMS II contains data for 406 detailed industries and 62 aggregate industries (BEA, 2013). RIMS II is the most transparent of the three models; it is essentially a set of tables, but it requires understanding of how to use the multipliers for economic impact analysis. The data sources are highly accessible, the tables provide a fairly disaggregated level of industry detail, and multipliers are updated to reflect recent wage, salary, and income data. However, limiting assumptions of the RIMS II model include the following: backward-looking linkages, fixed purchase patterns, industry homogeneity, no supply constraints, no regional feedback, and a static framework (BEA, 2013).

Impact Analysis for Planning

IMPLAN, which was initially developed by the U.S. Department of Agriculture Forest Service and has become a private company, is a more complex I-O model built around the concept of social accounting matrixes (SAMs). Similar to I-O tables, SAMs provide a double-entry bookkeeping system that can tracing monetary flows between industries (IMPLAN, undated). By definition, the SAM must balance (i.e., receipts must equal expenditures). For any industry to fulfill the demand for its goods and services with production outputs, it needs labor and the outputs of suppliers as inputs, generating further and indirect demand up the supply chain and in the labor market. IMPLAN's inter-industry models provide information on market transactions between industries and consumers and capture government transfers to individuals and businesses, tax payments by individuals and businesses, and transfers between individuals. SAMs expand on traditional

I-O tables by providing information on nonmarket financial flows (i.e., industry-institution transfers, inter-institution transfers).

IMPLAN consists of 534 industries, including private industries and government enterprises, which are more disaggregated relative to the BEA's 405 industries. Compared with RIMS II, IMPLAN makes data entry and model specification easier. The more user-friendly software package is more expensive than RIMS II and offers a formal modeling system with several capabilities that RIMS II does not have. IMPLAN also explicitly breaks impacts into direct, indirect, and induced effects. The limitations of IMPLAN are similar to RIMS II (generally similar to all I-O models), including that it is a static model of the regional economy. Users might also have to account for inflation if data are not from the current year. However, IMPLAN is much more customizable than the other available economic models because users can modify the models default assumptions.

REMI

REMI is the most complex and expensive of the three models but has capabilities that make it better suited for considering the implications of changes in consumer behavior or changes that occur over time. Specifically, REMI has capabilities that incorporate aspects of a model. Other proprietary models offer fully dynamic CGE models but are generally less transparent because of their proprietary nature. CGE models are generally more expensive to license than I-O models and require significantly more data inputs, time, and technical expertise.

Economic Models Used in Previous Standardized Regulatory Impact Assessments

As of December 2022, state agencies in California had conducted 77 SRIAs, excluding withdrawn and/or resubmitted regulations. Most SRIAs have used a standard set of economic tools to estimate the impacts of regulations. The scope of the regulation might help determine what macroeconomic model is most appropriate. Overall, the appropriate model choice for any SRIA depends on trade-offs between complexity, the level of industry disaggregation, transparency, and cost. There is no one-size-fits-all model. Instead, because these economic models have contrasting strengths and weaknesses, certain models might be best suited to certain analyses and types of regula-tions. Agencies might prefer models that vary depending on the relative eco-nomic magnitude and type of impacts a regulation is expected to generate.

As shown in Figure 5, the most widely used models are RIMS II (used by nine state agencies) IMPLAN (used by nine state agencies), and REMI (used by seven state agencies). Four agencies have also used other proprietary CGE-type economic models, which are less transparent but more customizable. Looking across all SRIAs, the REMI model was used most often (38 SRIAs). However, this finding is driven by a single agency (the Air Resources Board) that exclusively uses this model. The Air Resources Board used REMI used for 28 SRIAs, while other agencies used it ten SRIAs. In comparison, DIR has conducted seven SRIAs to date: three using IMPLAN and four using a proprietary CGE model. Agencies frequently work with contractors to conduct analyses using these various macroeconomic models. When state agencies in California first began preparing SRIA reports, some agencies

FIGURE 5

Number of California State Agencies Using Selected Economic Models, 2014 to 2022

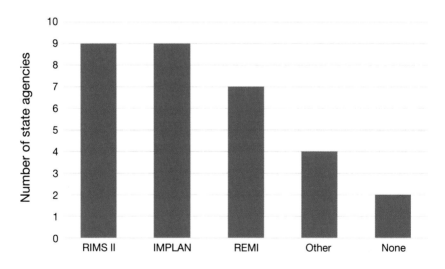

SOURCE: Features information from DOF, undated.
NOTE: Agencies might be counted more than once if they have used more than one approach.

failed to use any of the widely available economic models. However, these tools have become standard in state regulatory analyses.

Department of Finance Comments on Past Standardized Regulatory Impact Assessments

DOF acts as the initial reviewer of SRIAs submitted by state agencies for proposed major regulations. Once a SRIA is formally submitted, DOF publishes a comment letter within 30 days, to which the rulemaking agency must respond. Review of these public comment letters can help identify the most common methodological issues in conducting SRIAs and areas of particular concern for DOF. Figure 6 summarizes comments about the 77 SRIAs submitted to DOF between 2014 and 2022.

DOF most commonly noted shortcomings in meeting the following requirements:

- Describe the baseline reflecting the anticipated behavior of individuals and businesses in the absence of the proposed major regulation including existing laws and economic conditions.
- Address all fiscal impacts to state and local government agencies.
- Describe and evaluate the benefits and costs of at least two policy alternatives.
- Include analysis of the disparate impacts of the regulation.

DOF frequently provided additional technical guidance, including (1) describe qualitatively or attempt to quantify additional categories of benefits or costs that were not considered in the analysis, (2) provide the rationale for key assumptions or include a sensitivity analysis to show how impacts might vary under different scenarios, and (3) report total benefits and costs to individuals, businesses, and public entities on a disaggregated, year-by-year basis. The nature and context of the comments suggest that DOF pays particular attention to the fiscal effects of regulations on the state and local levels.

The nature and extent of DOF's comments has changed, which might reflect improvements in the quality and consistency of SRIAs and increased

FIGURE 6

Type and Number of Comments Given on 77 Standardized Regulatory Impact Assessments, Submitted from 2014 to 2022

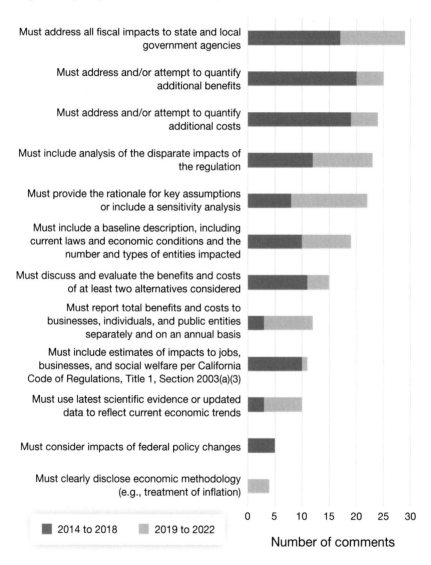

SOURCE: Features information from DOF, undated.

focus on assumptions, data sources, and methodology. Such change is evident when comparing comments given between 2014 and 2018 following the introduction of SRIAs in California compared with SRIAs submitted between 2019 and 2022. Generally, state agencies have become less likely to omit major analytic requirements, and DOF's comments have focused increasingly on transparency and justification of key assumptions (e.g., number and types of impacted individuals and businesses, anticipated policy responses) and methodology (e.g., selection of discount rates, treatment of inflation). For example, most of DOF's comments that recommended that agencies use the latest scientific evidence or updated data to reflect existing economic trends were issued between 2019 and 2022.

However, feedback from DOF continues to highlight some more persistent shortcomings. Figure 6 identifies the following as common recurring issues:

- inadequately describing the baseline, such as the number and types of affected individuals and businesses and potential policy interactions with existing (and enacted but not yet implemented) state and federal laws
- insufficiently addressing all fiscal impacts
- omitting distributional analysis of the impacts of the regulation.

Recommendations

After a review of guidance, analyses, and DOF's formal review of and comments on SRIAs developed between 2014 and 2022, we note that changes in California state administrative requirements for executive agencies have led to greater transparency in the rulemaking process for proposed major regulations and improvements in the quality and consistency of the economic analyses conducted. In evaluating the content, modeling choices, and feedback received on SRIAs developed by various state agencies, our objective was not to evaluate the overall regulatory process but to provide context to those who are now facing similar decisions. However, several observations arose naturally from the development of a guidance document for state agencies conducting SRIAs.

Therefore, we make several recommendations to ensure that regulatory analysis promotes rulemaking transparency and provides information to decisionmakers and the public to credibly identify policy actions that implement legislative objectives in the most cost-effective manner. These recommendations, particularly our second and third recommendations, are consistent with recommendations previously made by the Legislative Analyst's Office (Taylor, 2017). That these issues remain relevant suggests that many California state agencies and analysts are still learning the SRIA framework and requirements as well as issues pertaining the benefit-cost analysis more broadly. Furthermore, it suggests that existing guidance from DOF, codified in regulation and the State Administrative Manual, has left some gaps in terms of methodological approaches and best practices for conducting SRIAs. More-detailed standard guidance would help ensure consistent and justifiable approaches are used across regulations. We make the following recommendations:

- California state agencies should be careful to avoid common errors in developing SRIAs, such as those described in Figure 6. California state agencies should be in close communication with DOF for informal feedback in advance of the formal submission of a SRIA and should review available guidance materials for insights. Although this report is not an official source of guidance, it contains insights that agencies might find useful. We also include a general outline for a SRIA that addresses the major analytical requirements under the California Code of Regulations (Appendix B).
- The California legislature should direct DOF to develop more-detailed guidance on best practices for SRIAs and focus on common issues, such as describing the baseline, evaluating regulatory alternatives, assessing fiscal impacts, and identifying disparate impacts of regulations. DOF should also consider providing more-detailed guidelines on discount rates, the analytic time horizon, treatment of uncertainty, quantification and monetization of benefits and costs, and other areas in which it has frequently raised issue with assumptions in the analyses. In addition, DOF should consider providing more guidance on macroeconomic models and their strengths and weaknesses.

- The California legislature should consider removing certain analytical requirements that increase the rulemaking burden for state agencies yet provide limited informational value for decisionmaking. Estimating secondary macroeconomic impacts on jobs and businesses accounts for a significant portion of the SRIA analysis. State agencies must invest significant resources or contract with consultants to use economic modeling tools with specific capabilities. The modeling tools are complex, lack transparency, and rely on various assumptions about business and household behavior that are subject to uncertainty. Furthermore, none of the models are useful for capturing public health and safety effects, worker safety effects, environmental benefits, or the intangible benefits of regulations. In contrast, traditional benefit-cost analysis offers techniques to estimate monetary values associated with some of these outcomes. As demonstrated in both past federal regulatory analyses and SRIAs, agencies often make policy decisions using an evaluation of direct benefits and costs as a basis for rejecting regulatory alternatives. Additionally, the economic models described here cannot evaluate all impacts associated with the SRIA requirements, such as increases or decreases in the number of businesses in the state, which leaves agencies to provide qualitative discussions that have been less rigorously analyzed and might have little value for decisionmakers.
- The California legislature should direct DOF to require state agencies to publicly post updated analyses when economic methods, key assumptions, or findings in the SRIA are substantially changed after the document is initially reviewed by DOF or posted for public comment. A record of public comments and agency responses, including revisions to the economic analysis based on new information or data provided during the public comment period, should also be made public. Revisions to the economic analyses in the SRIA, including changes recommended by DOF, are documented in the Standard Form 399 as part of the final rulemaking package but might not be communicated to all policymakers or stakeholders. It is exceedingly rare for changes to the economic analyses to result in a revised SRIA. This shortcoming appears to undermine the goal of achieving rulemaking transparency. It is possible that agencies do not submit revised SRIAs because doing so would reset the administrative review process and delay the imple-

mentation of a proposed regulation by up to a year. The legislature should mitigate this practical concern by amending the APA to avoid updated analyses resetting the entire rulemaking process.

Conclusion

In this report, we describe the California regulatory process under the APA and provide guidelines for state agencies conducting regulatory analyses. SB 617 (California Senate, 2011) amended the APA to establish a new process for analyzing and reviewing major regulations. Specifically, the change requires that each state agency prepare a SRIA—a more detailed regulatory analysis—of any major regulation and submit it to DOF for review and comment before a rule is adopted.

The level of technical complexity, the methodological approaches, and general compliance with DOF's guidelines varies considerably across SRIAs. It is not surprising that the SRIAs completed as of this writing have been inconsistent, particularly initially, given that the requirements are still relatively new, only a tiny percentage of regulations are considered major regulations, and many agencies have only conducted one or two analyses as of this writing. The scope and level of specificity in the comments from DOF to agencies on the SRIAs have also evolved over time.

Rulemaking agencies can better navigate the rulemaking process by relying on guidance and other resources, such as consultation, provided by DOF and guidelines developed for federal regulatory impact analyses, such as OMB's Circular A-4 and the more comprehensive guidance developed by EPA (2014) and HHS (2016). To enhance evidence-based decisionmaking, rulemaking transparency, and administrative efficiency (e.g., during interagency review or public comment), agencies should effectively describe and explain the need for the proposed regulation, present feasible policy options for addressing the problem, carefully document the methodology and economic tools used in the analysis, evaluate the full variety of benefits and costs (as well as secondary macroeconomic impacts), and consider distributional effects across the affected population.

With these goals in mind, SRIAs are useful tools. State agencies must consider a wide variety of economic impacts, justify the need for a proposed regulation, and demonstrate that the regulatory approach is the most cost-effective policy option or achieves other stated statutory objectives. Generally, the quality and consistency of regulatory analyses in California has improved since state agencies began conducting SRIAs in 2014, but various methodological weaknesses and other analytical shortcomings remain. Agencies might benefit from more detailed guidance from DOF, and the California legislature might consider removing certain analytical requirements that increase the burden for state agencies but provide limited informational value for decisionmaking.

Since being introduced, SRIAs still present a challenge for state agencies in the collection and development of requisite data and information, scientific evidence, and expertise, and it is difficult for these agencies to devote the time to conduct such economic impact analyses. Nonetheless, the public SRIA reports provide a benefit to decisionmakers by identifying the most cost-effective policy options and to stakeholders by understanding the estimated benefits and costs, assumptions, and agency justifications for various rulemakings.

Development of an Occupational Health Standard

In Appendix A, we provide an overview of the regulatory process that is unique to the California Division of Occupational Safety and Health's (Cal/OSHA's) development of an occupational health standard. Figure A.1 shows the steps that Cal/OSHA follows in this process. Occupational safety standards are developed by the Occupational Safety and Health Standards Board and follow a similar process.

FIGURE A.1

Steps to Develop an Occupational Health Standard

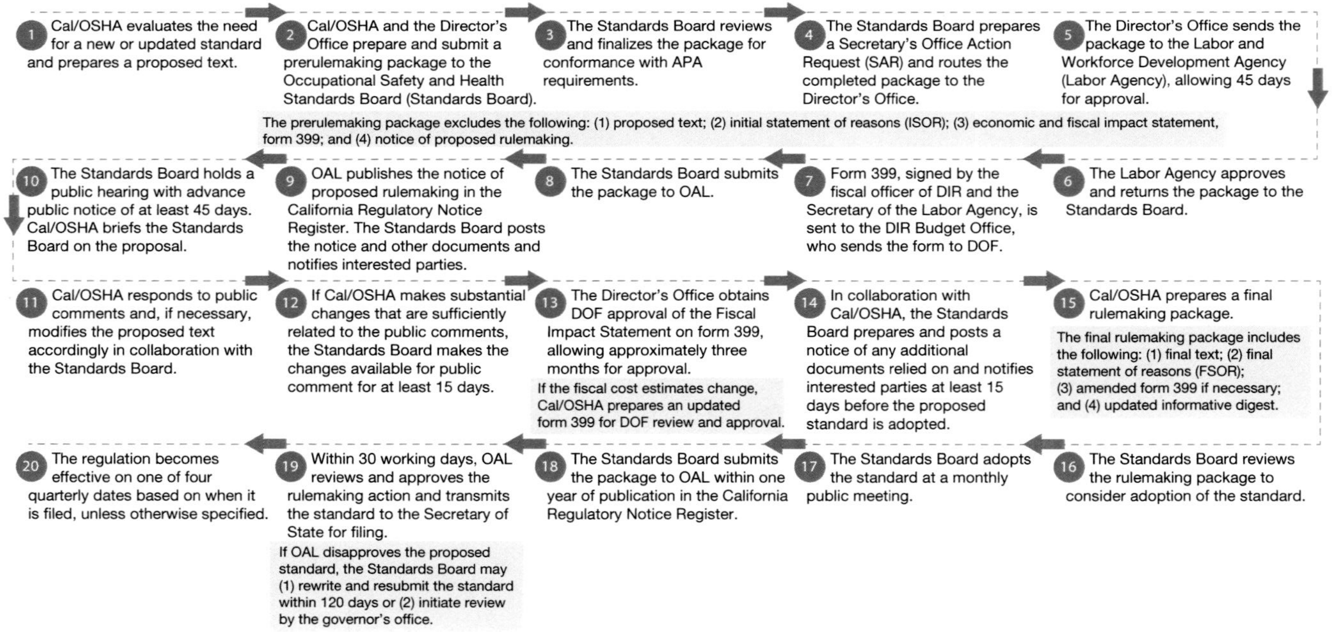

1. Cal/OSHA evaluates the need for a new or updated standard and prepares a proposed text.

2. Cal/OSHA and the Director's Office prepare and submit a prerulemaking package to the Occupational Safety and Health Standards Board (Standards Board).

3. The Standards Board reviews and finalizes the package for conformance with APA requirements.

4. The Standards Board prepares a Secretary's Office Action Request (SAR) and routes the completed package to the Director's Office.

5. The Director's Office sends the package to the Labor and Workforce Development Agency (Labor Agency), allowing 45 days for approval.

The prerulemaking package excludes the following: (1) proposed text; (2) initial statement of reasons (ISOR); (3) economic and fiscal impact statement, form 399; and (4) notice of proposed rulemaking.

6. The Labor Agency approves and returns the package to the Standards Board.

7. Form 399, signed by the fiscal officer of DIR and the Secretary of the Labor Agency, is sent to the DIR Budget Office, who sends the form to DOF.

8. The Standards Board submits the package to OAL.

9. OAL publishes the notice of proposed rulemaking in the California Regulatory Notice Register. The Standards Board posts the notice and other documents and notifies interested parties.

10. The Standards Board holds a public hearing with advance public notice of at least 45 days. Cal/OSHA briefs the Standards Board on the proposal.

11. Cal/OSHA responds to public comments and, if necessary, modifies the proposed text accordingly in collaboration with the Standards Board.

12. If Cal/OSHA makes substantial changes that are sufficiently related to the public comments, the Standards Board makes the changes available for public comment for at least 15 days.

13. The Director's Office obtains DOF approval of the Fiscal Impact Statement on form 399, allowing approximately three months for approval.

If the fiscal cost estimates change, Cal/OSHA prepares an updated form 399 for DOF review and approval.

14. In collaboration with Cal/OSHA, the Standards Board prepares and posts a notice of any additional documents relied on and notifies interested parties at least 15 days before the proposed standard is adopted.

15. Cal/OSHA prepares a final rulemaking package.

The final rulemaking package includes the following: (1) final text; (2) final statement of reasons (FSOR); (3) amended form 399 if necessary; and (4) updated informative digest.

16. The Standards Board reviews the rulemaking package to consider adoption of the standard.

17. The Standards Board adopts the standard at a monthly public meeting.

18. The Standards Board submits the package to OAL within one year of publication in the California Regulatory Notice Register.

19. Within 30 working days, OAL reviews and approves the rulemaking action and transmits the standard to the Secretary of State for filing.

If OAL disapproves the proposed standard, the Standards Board may (1) rewrite and resubmit the standard within 120 days or (2) initiate review by the governor's office.

20. The regulation becomes effective on one of four quarterly dates based on when it is filed, unless otherwise specified.

SOURCE: Features information from DIR, 2016.

General Standardized Regulatory Impact Assessment Template

In this appendix, we provide a general structural template for a SRIA that addresses the major analytical requirements under the California Code of Regulations. This approach is only a recommendation; agencies might want to tailor the report structure to the scope of their proposed regulations while addressing all of the required elements. For example, disparate impacts of the regulation might be discussed in a summary of direct costs and benefits.

1. Introduction
   a. Regulatory History
   b. Statement of the Need for the Proposed Regulation
   c. Overview of Proposed Regulatory Action
   d. Major Regulation Determination
   e. Baseline Information
   f. Public Outreach and Input

2. Direct Costs
   a. Direct Costs for Businesses
      i. Direct Costs to Typical Businesses
      ii. Direct Costs to Small Businesses
   b. Direct Costs to Individuals

3. Direct Benefits
   a. Direct Benefits to Businesses
      i. Direct Benefits to Typical Businesses
      ii. Direct Benefits to Small Businesses

b. Direct Benefits to Individuals

4. Fiscal Impacts
   a. Fiscal Impacts on Local Government
   b. Fiscal Impacts on State Government

5. Macroeconomic Impacts
   a. Methods and Model Selection
   b. Inputs and Assumptions
   c. Results of the Macroeconomic Assessment
      i. Creation or Elimination of Jobs in California
      ii. Creation of New Businesses or the Elimination of Existing Businesses within the State
      iii. Competitive Advantages or Disadvantages for California Businesses
      iv. Increase or Decrease of Investment in California

   d. Incentives for Innovation in Products, Materials, or Processes

6. Regulatory Alternatives
   a. Alternative 1
      i. Costs
      ii. Benefits
      iii. Comparison with the Proposed Regulation
      iv. Reason for Rejecting

   b. Alternative 2
      i. Costs
      ii. Benefits
      iii. Comparison with the Proposed Regulation
      iv. Reason for Rejecting

7. Distributional Effects of the Regulation
8. Conclusion or Summary of Economic Results

Abbreviations

APA	Administrative Procedure Act
BEA	U.S. Bureau of Economic Analysis
BLS	U.S. Bureau of Labor Statistics
Cal/OSHA	California Division of Occupational Safety and Health
CGE	computable general equilibrium
DIR	California Department of Industrial Relations
DOF	California Department of Finance
EO	executive order
EPA	U.S. Environmental Protection Agency
HHS	U.S. Department of Health and Human Services
IMPLAN	Impact Analysis for Planning
I-O	input-output
OAL	California Office of Administrative Law
OMB	U.S. Office of Management and Budget
REMI	Regional Economic Models, Inc.
RIMS II	Regional Input-Output Modeling System
SAM	social accounting matrix
SB	Senate Bill
SRIA	standardized regulatory impact assessment

References

Ballotpedia, "Agency Dynamics: States That Require Administrative Agencies to Conduct Cost-Benefit Analysis Before Implementing Rules," webpage, undated. As of November 22, 2023:
https://ballotpedia.org/Agency_dynamics:_States_that_require_
administrative_agencies_to_conduct_cost-benefit_analysis_before_
implementing_rules

Baxter, Jennifer R., Lisa A. Robinson, and James K. Hammitt, *Valuing Time in U.S. Department of Health and Human Services Regulatory Impact Analyses: Conceptual Framework and Best Practices*, U.S. Department of Health and Human Services, 2017.

BEA—*See* U.S. Bureau of Economic Analysis.

BLS—*See* U.S. Bureau of Labor Statistics.

California Code of Regulations, Title 1, Division 1, Chapter 2, Underground Regulations.

California Code of Regulations, Title 1, Division 3, Chapter 1, Standardized Regulatory Impact Assessment for Major Regulations.

California Department of Finance, "Major Regulations SRIAs and Calendar," webpage, undated. As of November 22, 2023:
https://dof.ca.gov/forecasting/economics/major-regulations/
major-regulations-table/

California Department of General Services, *State Administrative Manual*, June 2014.

California Department of Human Resources, "Pay Scales," webpage, last updated June 18, 2020. As of November 22, 2023:
https://www.calhr.ca.gov/state-hr-professionals/Pages/pay-scales.aspx

California Department of Industrial Relations, "Steps to Develop an Occupational Health Standard," webpage, last updated November 2016. As of November 22, 2023:
https://www.dir.ca.gov/dosh/steps-to-develop-an-ohs.html

California Government Code, Chapter 3.5, California Administrative Procedure Act, Sections 11340-11365.

California Office of Administrative Law, "About the Office of Administrative Law," webpage, undated-a. As of November 22, 2023:
https://oal.ca.gov/about-the-office-of-administrative-law

California Office of Administrative Law, "About the Regular Rulemaking Process," webpage, undated-b. As of November 22, 2023:
https://oal.ca.gov/rulemaking_participation

California Office of Administrative Law, "Administrative Procedure Act & OAL Regulations," webpage, undated-c. As of November 22, 2023:
https://oal.ca.gov/publications/administrative_procedure_act/

California Senate, State Government: Financial and Administrative Accountability, Bill 617, Chapter 496, October 5, 2011.

Cooke, Roger M., *Experts in Uncertainty: Opinion and Subjective Probability in Science*, Oxford University Press, 1991.

DOF—*See* California Department of Finance.

EO—*See* Executive Order.

EPA—*See* U.S. Environmental Protection Agency.

Executive Order 12291, "Federal Regulations," Executive Office of the President, February 17, 1981.

Executive Order 12866, "Regulatory Planning and Review," Executive Office of the President, September 30, 1993.

HHS—*See* U.S. Department of Health and Human Services.

IMPLAN, "Introducing the SAM," webpage, undated. As of November 22, 2023:
https://support.implan.com/hc/en-us/
articles/115009674708-Introducing-the-SAM

Interagency Working Group on Social Cost of Greenhouse Gases, *Technical Support Document: Social Cost of Carbon, Methane, and Nitrous Oxide: Interim Estimates Under Executive Order 13990*, February 2021.

Morgan, M. Granger, and Max Henrion, *Uncertainty: A Guide to Dealing with Uncertainty in Quantitative Risk and Policy Analysis*, Cambridge University Press, 1990.

OAL—*See* California Office of Administrative Law.

OMB—*See* U.S. Office of Management and Budget.

O'Hagan, Anthony, Caitlin E. Buck, Alireza Daneshkhah, J. Richard Eiser, Paul H. Garthwaite, David J. Jenkinson, Jeremy E. Oakley, and Tim Rakow, *Uncertain Judgements: Eliciting Experts' Probabilities*, John Wiley & Sons, 2006.

Organisation for Economic Co-operation and Development, "The Costs and Benefits of Regulating Chemicals," webpage, undated. As of November 22, 2023:
https://www.oecd.org/environment/tools-evaluation/costs-benefits-chemicals-regulation.htm

Public Law 96-354, Regulatory Flexibility Act, September 19, 1980.

Public Law 104-121, Contract with America Advancement Act of 1996, Section 201, Small Business Regulatory Enforcement Fairness Act, March 29, 1996.

Public Law 111-203, Dodd-Frank Wall Street Reform and Consumer Protection Act, July 21, 2010.

Public Law 111-240, Small Business Jobs Act of 2010, September 27, 2010.

Robinson, Lisa A., James K. Hammitt, Michele Cecchini, Kalipso Chalkidou, Karl Claxton, Maureen Cropper, Patrick Hoang-Vu Eozenou, David de Ferranti, Anil B. Deolalikar, Frederico Guanais, Dean T. Jamison, Soonman Kwon, Jeremy A. Lauer, Lucy O'Keeffe, Damian Walker, Dale Whittington, Thomas Wilkinson, David Wilson, and Brad Wong, *Reference Case Guidelines for Benefit-Cost Analysis in Global Health and Development*, May 2019.

Robinson, Lisa A., James K. Hammitt, and Richard J. Zeckhauser, "Attention to Distribution in U.S. Regulatory Analysis," *Review of Environmental Economics and Policy*, Vol. 10, No. 2, Summer 2016.

Steuerle, Eugene, and Leigh Miles Jackson, eds., *Advancing the Power of Economic Evidence to Inform Investments in Children, Youth, and Families*, Committee on the Use of Economic Evidence to Inform Investments in Children, Youth, and Families, National Academies Press, 2016.

Taylor, Mac, *Improving California's Regulatory Analysis*, California Legislative Analyst's Office, February 2017.

U.S. Bureau of Economic Analysis, *RIMS II User's Guide: An Essential Tool for Regional Developers and Planners*, December 2013.

U.S. Bureau of Labor Statistics, "Employer Costs for Employee Compensation," webpage, undated. As of November 22, 2023:
https://www.bls.gov/ecec/home.htm

U.S. Bureau of Labor Statistics, "Occupational Employment and Wage Statistics: May 2022 State Occupational Employment and Wage Estimates, California," webpage, last updated April 25, 2023. As of November 22, 2023:
https://www.bls.gov/oes/current/oes_ca.htm

U.S. Census Bureau, "Current Population Survey (CPS)," webpage, last updated October 20, 2023. As of November 22, 2023:
https://www.census.gov/programs-surveys/cps.html

U.S. Department of Health and Human Services, *Guidelines for Regulatory Impact Analysis*, 2016.

U.S. Department of Health and Human Services, *Guidelines for Regulatory Impact Analysis, Appendix D: Updating Value per Statistical Life (VSL) Estimates for Inflation and Changes in Real Income*, 2021.

U.S. Department of Labor, *Labor Cost Inputs Used in the Employee Benefits Security Administration, Office of Policy and Research's Regulatory Impact Analyses and Paperwork Reduction Act Burden Calculations*, 2016.

U.S. Department of Transportation, *The Value of Travel Time Savings: Departmental Guidance for Conducting Economic Evaluations Revision 2 (2016 Update)*, 2016.

U.S. Department of Transportation, *Treatment of the Value of Preventing Fatalities and Injuries in Preparing Economic Analyses*, 2021.

U.S. Environmental Protection Agency, *Guidelines for Preparing Economic Analyses*, 2014.

U.S. Environmental Protection Agency, *Handbook on Valuing Changes in Time Use Induced by Regulatory Requirements and Other EPA Actions*, 2020.

U.S. Environmental Protection Agency, *Report on the Social Cost of Greenhouse Gases: Estimates Incorporating Recent Scientific Advances*, 2023.

U.S. Office of Management and Budget, *Circular A-4: Regulatory Analysis*, September 17, 2003.

U.S. Office of Management and Budget, *Circular A-4: Regulatory Analysis*, November 9, 2023.